Living Green

Recycling

By Meg Gaertner

level 2 little blue readers

www.littlebluehousebooks.com

Little Blue House is distributed by North Star Editions:
sales@northstareditions.com | 888-417-0195

Produced for Little Blue House by Red Line Editorial.

Photographs ©: Shutterstock Images, cover, 4, 7, 11, 12 (top), 12 (bottom), 15 (top), 15 (bottom), 16–17, 18, 21, 23, 24 (top left), 24 (top right), 24 (bottom left), 24 (bottom right); iStockphoto, 8–9

Library of Congress Control Number: 2022901946

ISBN
978-1-64619-597-8 (hardcover)
978-1-64619-624-1 (paperback)
978-1-64619-675-3 (ebook pdf)
978-1-64619-651-7 (hosted ebook)

Printed in the United States of America
Mankato, MN
082022

About the Author

Meg Gaertner enjoys reading, writing, dancing, and being outside. She lives in Minnesota.

Table of Contents

Too Much Waste

People create a lot of waste. They use different goods. Then they throw those goods away.

A girl gets a cardboard box in the mail.

She opens the box.

She takes everything out.

She throws the box away.

cardboard box

A boy drinks from a can.
He empties the can.
Then he throws the
can away.

can

Cities collect waste.
Waste goes to landfills,
but it is not gone.
Waste stays in landfills for
a very long time.

landfill

wood

metal

Recycling

Goods are made of different materials. Paper is made from wood. Cans are made from metal.

People can use some materials again. They can make new items out of the old materials. This is called recycling.

Recycling reduces the amount of waste. Less waste goes to landfills.

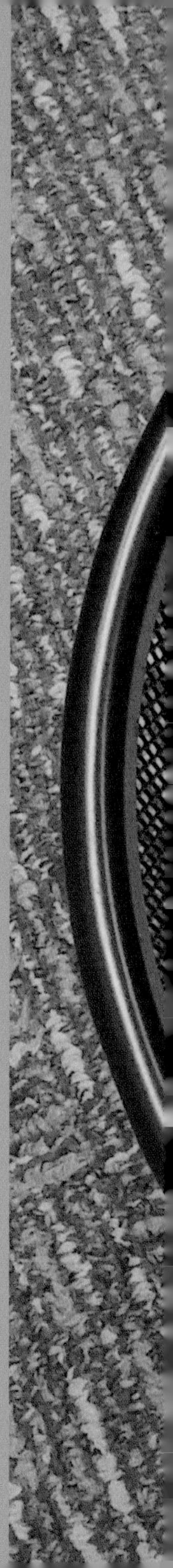

cardboard
paper
metal

How It Works

Certain materials can be recycled. Do not throw away paper, metal, or cardboard. Recycle them instead.

Use recycling bins.

These bins are often blue.

They have symbols.

The symbols show
three arrows.

recycling bin
symbol

City trucks take the bins. People or machines sort and clean the materials. Then machines recycle them.

truck

Glossary

landfill

recycling bin

metal

truck

Index